## Dedication

*Completion of this book can only be said to have been possible because of all the friends for 40 years as a member of Lotus Corps. This group encourages and inspires to love the Lotus Marque and all owners thereof. I could name individuals who have done more but the list would then just become the membership in its entirety.*

*Even more thanks to my partner for life, Jackie who married me over 50 years ago, and joined with me to embrace Lotus.*

To view photos in full color visit Lotuscorps.org plus other pictures not in this book.

# Europa *del Sol*

A blow by blow account of a
seven years war
Restoration of a classic Lotus

By
William Greenwald

**Europa Resurrection**                    **Part 1**


Xxx"It was a cold and stormy night"-- Really I don't remember what kind of night it was. But I do remember someone was writing a novel and was trying to start off with the above line. Please help me out here. Anyway I was doing my normal checking out of Loti on the internet for sale, e-baying as I do for parts etc. You may or not know the Lotus Seven of mine was also an e-bay purchase. That's three out of four Lotus's bought at auction by yours truly. For kicks I started to bid as this was the true resurrection I was looking for to test my talents as a sports-car restorer. Hell, a few brews, screwdriver, spray-gun, and two or three months and were done.

To make a long story even longer my sought after prize was mine for the taking at dollar amount I could live with. Now to tell the wife of this "treasure", could she live with my steal? ("Treasure" is a plum from Sue Herzog). I forget what I bribed her with but it must have worked as we are still married. I know when we got the Seven it was a life crisis that by not having owned a Seven: I would go to my just rewards, unrewarded. Somehow I got away with that one.

Next having this treasure in hand, monetarily, I needed to physically procure the beast. Terms of the auction were "must be picked up in seven days". Wow what great timing, the snows of December, 2005 were upon us and traveling to the east coast should be a great adventure. Logistics required I borrow the trusty Herzog trailer, a tow vehicle and a good road map, which my co-pilot cannot read or fold. I put in a call to Zog and made my request to borrow the two-wheeler. Bob asked where this 1970 Europa S2 was located and my reply was "Cape Cod" to which Bob immediately told me to hold the line. Early that Saturday morning, Bob came back asking if I would in turn deliver a car on my way to Boston, an Elva Courier he had sold but not yet delivered. The buyer, believe it or not, was less than 15 miles from where I was to retrieve the Europa. The daughter-in-laws SUV Mercury is equipped with a tow hitch, lights, etc and was available, so I got a new ball, wiring harness and put Jackie and a six pack in the car, six pack being, snow brush, diet Coke, lock de-icer, toll money, credit card, Elva parts, and an Elva in tow. We made good time and only needed to stop for gas, food, and to knock the ice and snow from the trailer.

We arrived in Massachusetts at the appointed place to find in my haste to leave early to beat the projected snowstorms, several messages were on my answering machine, after we left, that party accepting the Elva would not be available, being in sunnier climes. His wife and mother, plus a neighbor were able to assist in the unloading and stashing into his garage with others of the Elva persuasion. I'm sure before we even left the wife was calling her lawyer to start divorce proceedings, she weren't happy; cold, snow, middle of the night and another Elva! Jackie and I returned on our journey, another 15 miles. Some two hours later, we arrived with a remembrance of my army days, stationed outside of Boston. When asking directions, there is a saying in New England with unmarked roads, poor street lighting, rotary's, which is, "You Can't Get There From Here!" believe me it is true, at least nearly true. The owner of the Europa was standing in the middle of the road signaling me with a flashlight, my code is pretty poor but I assume it was not complementary. After an exchange of pleasantries we loaded the Europa onto the trailer, many parts into the SUV rear hatch. We then secured the spare frame to the side of the trailer with the obligatory red and yellow streamers. Transfer of funds, the title and manual, instructions to the Thru-Way and off into the night trying to avoid the forecast of a severe snowstorm approaching from Canada and points west. It took some doing but we managed to leave the beautiful state of Mass. by 1:30 am.

One cool thing about tooling around the country in the dead of winter with two strange sports cars it brings out many questioners as to "What in the Hell is That" and many "Boy I remember that from way back when or I had one of them in the 60's", so we had a few conversations along the way and maybe a few more people shaking their heads??? At any rate we arrived safely back in Lyons with the "treasure", having spent almost as much on travel, sightsee, transport, and snow-skid as the purchase price of the car.

Elva Going

---- Lotus Coming

Other Lotus Cars rebuilt by the author over the years.

BILLS 7 -- The author's 1972 Lotus Seven Mk 4 was a less extensive project car before del Sol. Preceded by two other Europa models. Finish color after black was French Blue.

.LOTUS 46 – Second in the Europa series is the Series One Type 46 hence the license tag LOTUS 46. Started white was repainted to BRG (British Racing Green) by author.

Xxx"It was a cold and stormy night"…---

I'm still trying to use that line, maybe for my murder mystery. Which by the way this may turn into, once my wife got a good look at the "treasure" I believe that is what she wanted to do to me, all bets are still not off. We stuffed the Europa trailer and all onto the driveway with a shoehorn and the last few feet into my carport with two more shoehorns. Unloaded the remainder of treasure trove and settled in for a long winter's nap, (another story, the night before Santa).

Lotus arrives in Lyons time to shoehorn in drive.

No longer a cold and stormy night and by the light of day, I investigated the totality of the "treasure". Rust is the order of the day on any exposed, or not, steel surface; frame, suspension, engine, gauges. Followed closely by peeling paint, broken and cracked fiberglass, demoralized dashboard, failing door hinges. Eureka! I found my total project car and decided it may take a few more weeks than the original estimate and maybe a few more tools, parts and beer!

Once the seats were removed, an easy task, just lift and out they come. Careful don't disturb rust and rot. I discovered what the firewall behind the cockpit was made of, "nothing," there is not to be found. Gives great access to the water pump, pulley, belt, if this were a twin-cam version. Being a Renault this had no great advantage. The only thing more plentiful than rust were acorns and exhibits of small animals setting up housekeeping inside of various parts of the Europa cabin, frame, basically every nook and cranny (I could not find these parts in manual. but I've heard of them).

Let's take it easy step by step with baby steps first. Clean and clear. Watch words of the day. I started with easy flat and very paint peeling surfaces. The front and rear deck lids. Whoever repainted the car, red, did so without proper surface preparation. This is a key step in any Lotus restoration or repair, get down to the glass; I found two other colors, white and green. I believe the factory color was BRG. The tool I found worked the best was a single edge razor window scraper. The paint just jumped off the car with little dust and usual clouds when using a sander or other power tool. Only to find copious amounts of Bondo! Hidden under the detaching paint trying to reshape the lids to something other than original specs, it seems they wanted a flat surface at the edges which is not stock. Of course the inlet screens on the rear deck lid are trash, they are steel, and I am beginning to think there was a flood in this area of the country during storage, ala Noah's Ark, with rust from the flood and rain, acorns for feed and straw/fur for animal habitats.

Easy, lets move on to something more complicated. The DOORS, not the rock group, but the ingress, egress devices. Anyone familiar with the hinges on Europas knows a damp day will prevent removal of the hinges any time after two weeks of original car build date. Can you imagine almost 40 years. Both doors were cut from their openings using a hack saw and much swearing. The ½ inch shafts are not easy to get to and stroke is limited and I broke too many blades on the saber and Sawsall tools to do it with power. Normally the hinge can be cut inside the door shell and lowered or raised from the bobbins. On this car the shafts were not letting go of the bobbins, so cutting outside between the door and body was necessary. Once out I found both left and right bottom door pockets broken out that allowed the removal of the steel bobbins which constrain the hinge shafts. With all the genius of Lotus Engineering you would have thunk a better system would have been proposed. On further inspection it was discovered a further repair guffaw was evident on the passenger door. Approximately 6 to 8 pounds of putty, Bondo and fiberglass were wrapped around the door hinge. The only way to extricate the material was to perform major surgery on the door shell cutting the hinge portion out to expose the cancer. Every time I find another problem solved by an unknowing or uncaring clod I could spit, I'm being nice now. If anything we are fortunate in Lotus Corps; we have expertise in all aspects of our cars construction and repair, so anyone can plead ignorance, I'm #1, and ask for help from club members who like economists will point in all directions at the same time eventually giving the correct advice and procedure to follow.

Still using the scrapper and sanding technique I was able to strip down the doors to arrive at respectable surfaces inside and out. All the time the mind is working on what modifications could be made to improve operation and appearance. During this experience I have found to put down on paper an idea or method, think it over again and again, put it on the side and come back later, sober and rethink. Before each endeavor I found at least one or two ways to approach better than the first potentially catastrophic method, "Patience" is another watchword in restoration, (schedule for completion now past 4 months). You must look for an easier, more positive approach, watch out for pitfalls such as putting two things in the same place or not being able to put something back in. For instance I had considered glassing in the top access hole in the doors, not good if you want to reach in to the door handle mechanism. Why would I want to close in the hole for the side window, simple I don't want side windows with heavy motor operator, guides and glass? Aha! We are on our way to making a Europa Convertible, next time, "Europa del Sol"!

Europa with see through firewall; note copious amounts of rust everywhere.

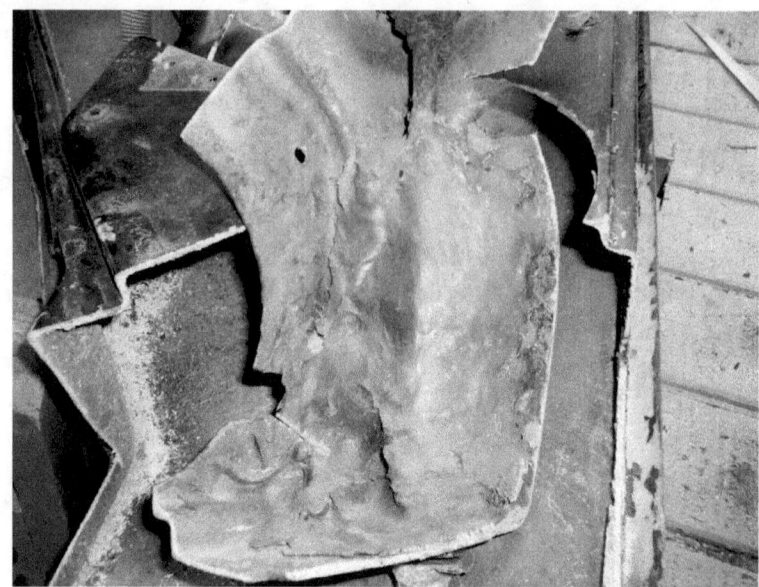

Somewhere under the blob of resins, etc lives the notorious rust ladened door hinge pin. Hard to see but bobbin is right above the word "resins" and the shaft is straight up due north.

Start of dismantling process, trash city dashboard, cockpit.

Xxx"It was a cold and stormy night"...---

Doors are out, most paint removed to expose problem areas of spider cracks and broken glass. Time to remove all of the hardware; hinges, window glass and frame, and power window motor operator all items heavy, remind me to put on the scale someday to find out exact weight savings. This is about time I decide to make the Europa Coupe into a convertible. "Why"? You may ask, my first reply is "Because I Can and it's Here" same as asking why people climb mountains. (Poetic license) My apologies to Carl and other Europa purist.

Anyone who has owned or driven a Europa in the summer knows how hot they can be. And except for the Series 1 which had pop-out windows of plexi-glass the later versions with power windows went only three quarters of the way down admitting little to no air, and the blower would only transfer super heated air from the front compartment. So out with the windows, weight and all and "Off with their Heads", (Quote the Queen of Hearts), I know of her because I met Alice in Wonderland most of the time I live there. Top and side of the frame must go too, leaving only the wing glass for better air flow.

I have seen and heard of highly modified Europa(s) a few with tops and windows gone totally for racing. Seems the top is a strong part of body stiffness. Even though everyone is calling it "del Sol", I actually had this design idea for many years way back to my Series 1 days, (before Honda) keeping the back window frame for stiffness and rollover bar and good looks. Oh Yeah!, I forgot the back window glass is a goner too, flow thru ventilation, and great engine sounds behind the head. Did I mention this will be only a Fair Weather vehicle?

Seeing as I purchased a welder, MIG, to work on seats, frame, etc. I decided to provide a measure of stiffness to the body. Again I cannot stress enough the need to plan and scheme everything out well in advance, I gave the 1" square tubing about 5 minutes to start the top most rail just under the roof line. The remainder took a few more days and thought process to arrive at the design pictured. As Lotus and Colin Chapman always planned, use any component for more than one function, my frame multi-tasks too. The entire tubing frame will attach to the steel backbone chassis of the Europa. The Series Two has 3 bolts behind the seats on either side of the center tunnel which attach to a sheet metal angle of the chassis. I started with steel plates drilled and mounted to the six holes. Followed the profile of the body and seats, with square tubing, to just below the window where another square tube runs from side to side connecting the top rail and side rails that end below the door sills and bolt to the outer seat belt retainer brackets. So the pseudo-roll bar; provides stiffening, roll over protection, attachment point for seat belts (lap and shoulder) and provides a frame for the fiber-glassing in of the openings. The frame also provided a guide for the Sawsall, cutting the roof off. The weight savings of the door windows, etc. will probably equal the weight of the heavy wall welded tubing and extra fiberglass. Because the firewall was 80-90% shot I cut out the openings behind the seat to square off the ports, keeping as much of the double wall fiberglass body panels, and provide access for the fuel tanks, newer 7 gallon tanks on each side. Finally the steel framework provides a mounting surface for closure and insulation mats and trim pieces later in plan. Remember plan ahead and use every piece to its fullest potential.

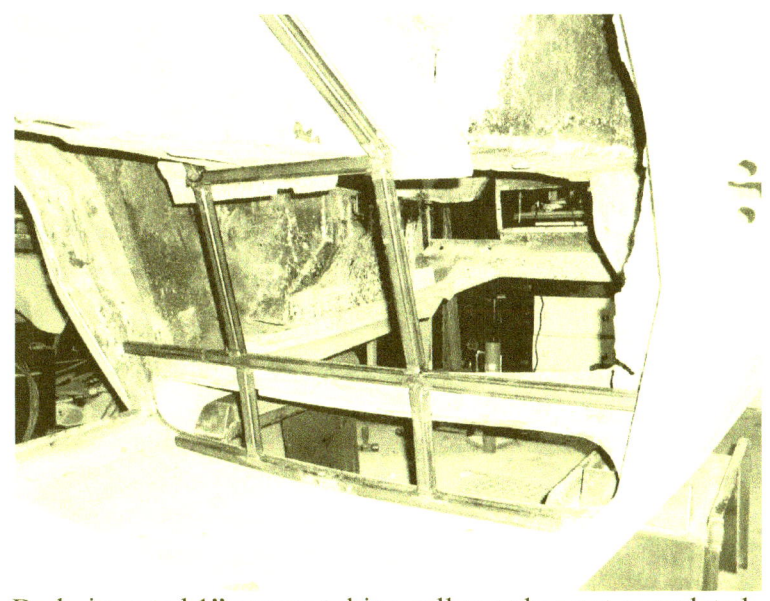

Body inverted 1" square tubing roll over bar not completed.

Door pillar and roof section gone.
Roll bar complete but not glassed in.
Please ignore the air scoop behind the door we have not got there yet by text description.

Xxx"It was a ~~cold and~~ stormy night"...---

Last month reported on the roll over frame. Even though it wasn't next we shall continue on for continuity sake. When I figure out exactly how; I will place the photo album on the web, but for now know there are some 300 photos taken on this project. Keep a running log if only to remind yourself later what came off of where and how bad did it really start out, or if you desire to compose a missive.

The front portion of the roof removal entails keeping the windscreen frame intact. A trick I learned was a method of filling in and spacing using nylon rope, other materials will do, because it has some strength forms well and keeps spacing thickness. For the roof portion I wanted to keep, I added three ½" diameter rope sections glued in place to the underside of the existing roof. Once in place two layers of glass fiber mat were added and resin added. The contour line was followed with Sawsall cutting the center segment of the fiberglass roof, the rope in front and square tubing at rear window. After removing the edges were formed into the body with additional layers of glass and resin.

To add strength to the windscreen frame the side pillars have a valley on the inside which was an ideal fit for a ½" diameter stainless steel bar placed in a bed of fiberglass and covered over completely filling in the valley, top of windscreen to the area of the door post under the dash. This should provide a stiff form into which the standard or plexi windshield can be fixed. (Haven't decided yet). Of course all of this glass work needs to have a final finish as it will be exposed and match the exterior/interior scheme.

The stock Europa has a black leather(ette) trim and sun visor mounted off the side of roof panel. These pieces are out of picture now. Left to add a method of attaching a toneau cover roof to the windscreen frame, well in the future, there are three plans for this, later.

The rear roof section with the 1" square tubing on underside of roof line needed to be enclosed in glass as well as sides and other exposed steel framework of the roll over bar. This makes the frame an integral part of the body and dresses up the Targa section with smooth lines ready for paint.

Three rows of rope glassed into roof, stainless rod on side pillar. Body inverted.

Top center section removed; note all of this work done with body upside-down. Easier to layer glass and not work overhead.

Side pillar of windscreen frame with stainless bar installed ready to glass in.

Xxx"It was a ~~cold and stormy night~~"... —

Work continues on the roof removal and subsequent glassing in of steel framework. Bad planning here: too much vertical work, hard to layer up glass with ease of a horizontal surface. I struggle through doing small sections at a time.

Next brain storm again a throw back to when I had planned to make the Series 1 wider in the rear. When autocrossing the Europa more tire is normally a good thing. Consider the stock Series 1 had 135 x 70 rock hard radials (Dunlop SP) from the factory, with the optional alloys sporting 145's. My last tire selection on "Kermit" was 205 x 60 rear and 185 up front. Yokohama 008rs for grip and handling way better. However, they did rub on the fender lips. The lip in question on the Europa is a horizontal glass piece sticking inside after the gentle roll of the approximate 1" radius. This adds to strength and is a very nice form great flow of material. So to prevent the rub out comes the grinder to remove as much material as possible front and rear wheel wells lips to eliminate the rub. Pretty good job on rear but hard turns under braking did not quite cut it up front. Many racers opt for fender flares; I however preferred to make the rear end wider by 3 inches. Actually started out as 3 per side for 6 inches, didn't look right, need to be flexible and allow for such events.

When sanding down the body to remove the red, green and whatever paint I found a rather thin segment of fiberglass. This thin was so thin I was plunging through the skin with hand sanding. I do very little with power tools. This line happens to be where the "sail" or top of fender, behind doors, contacts the lower bowed out section that starts out with a large radius to almost vertical lower portion of the fenders. How fortunate this is the exact line I had intended to follow on the Series 1 and now on this car the opportunity presented itself with the ability to follow this line with only a razor knife for a straight smooth cut from 3" behind the door to the rear bumper. "KISS" is a formula I try to follow. "Keep It Simple Stupid". The line I followed had to be simple, don't get into the double wall section of wheel well, don't get into curves in rear section, and don't make trunk lid or opening wider. Keep mounting to chassis the same and most important keep location true to body line. Another item I am not fond of on the Europa are the chrome bumpers. They were gone on Kermit and are planned to be gone on del Sol. So making the rear wider through the bumper line presents no problem. As can be seen in the photos the line I opted for met my KISS plan. Cutting the rear between the grill and tail lights, splitting the lower cowl behind the joint section and splitting the joint section on the underside of the car behind the wheel well. Simple!

For alignment I made up straps 2 inches wide by 6 inches long of 16 gauge steel with a series of holes on either side of the split line for ¼" bolts. Drilled a second set of holes 1-1/2" wider on one side these were then bolted to the body, before cutting the section off, for horizontal and vertical alignment. A couple at the back/bottom of shell, two in the back wheel well and another at front of wheel well. Once the cut is made simply move the bolts to the second set of holes and all is true and ready to re-glass section to body. Whoa! More items to consider don't start mixing the resin, yet!

Several bad section of glass on the lower fender cracked and need repair, likewise portions of the rear bumper and tail lights and grill opening and chassis mounting and , and and. Need to fix all of the ands first while access to sections is easier and/or can be rotated to a horizontal position for layups. Sanding and paint removal on body is still not complete so some of this needs doing. I think the project may now be closer to one year to complete.

Who am I kidding let's make that two years or more. We haven't even gotten into the rear mount radiator and side air scoops yet, save that for next time.

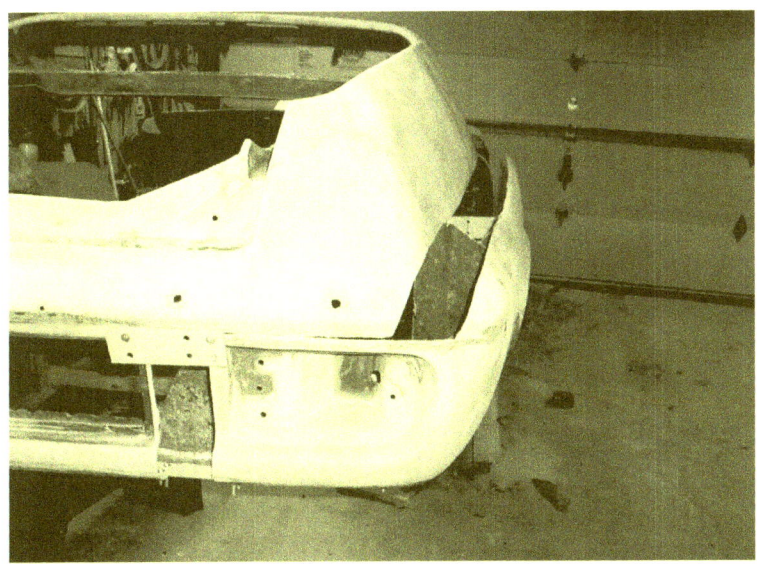

Rear section at 3" wide later revised to 1-1/2" Note brackets and cut line followed.

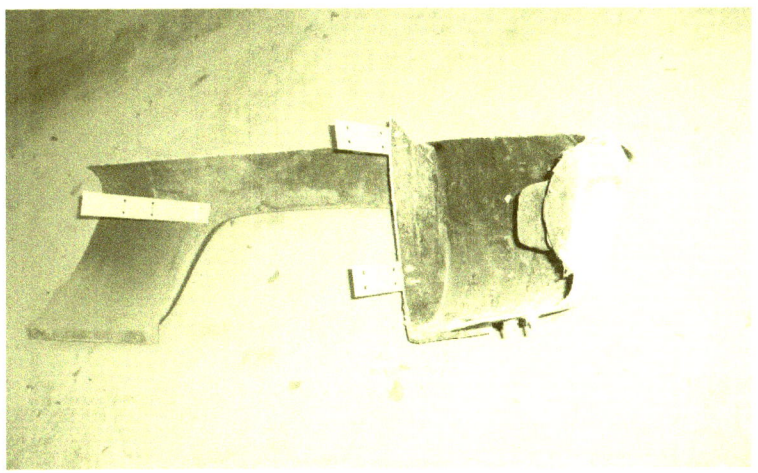

Note simple cut lines only one through the wheel well. Front portion of wheel well is closure piece for access to door latch

is a composite material. The only difficult cut line is leading edge for air scoop. Here we just followed the door contour at a fixed distance behind the door opening. Also note the brackets and two series of mounting spacing holes.

First test fit of the side panel widening air scoop location.

Before attaching the rear sections cut from each side of the Europa, at the required 1-1/2 inch spacing, consideration needed to be given to strength of the fiberglass panels. Remembering back to removal this is very thin glass panel which I cut through with a razor knife. I again used the nylon rope concept to add thickness of a rolled over edge on both the interior and exterior edges of the air scoop openings. The plan of course is to do this and all other glass repairs before installing the sections to first, get to the panels and second, keeping the surfaces horizontal for ease of lamination of fiberglass.

Following the contour of the cut panel the rope is glued to the panel at a reasonable distance from the edge allowing for a build-up of mat. The sections are heavy enough to hold shape and no stress was put on otherwise when installed on car the opening could be irregular. I checked this first with a cardboard cutout of the fender curvature to assure correctness. Once the rope was glassed in other repairs to any cracked or broken pieces were accomplished. It is important to note before any of the work was done all areas to be glassed were cleaned of dirt and oil to insure proper adhesion of the laminate.

The vertical interior surface behind door opening was more difficult to mold in the fiberglass and rope. Using multiple layers and less resin (dryer) the glass will stay in place, another tip from Paul Quiniff, Fiberglass Solutions.

Paul and I had several discussions on the method and profile of attaching side panels back onto the car. Granted we have good spacing and alignment but how to get a good body line without resorting to complicated molds. All of his ideas and direction seemed too complicated to me so again I resorted to a long process of brain-storming, plan and re-plan. Draw some pictures in mind and on paper. Remember patience! My final plan was to mold segments of a "Z" shaped section which would be riveted to the inside of the top and bottom separated sections of the fender panel. A simple aluminum sheet was bent to the approximate form needed. Wax paper was used as a parting agent to the molded pieces which consisted of two layers of fiberglass mat. Because the profile of the side panel is a complex curve the segments were made in various lengths of 2-6 inches to allow their straight profile to match the curve without a complicated mold. The mold allowed for about 24 inches of material so several applications were required. At the same time other areas such as fender-well needed straight section of flat sheet. I again resorted to an aluminum sheet and again formed two layers of glass wide enough to close the gap of 1-1/2 inches plus allowing for pop-riveting to either side.

Once the molds cured and fender sections ready to re-install a series of holes for 1/8" pop rivets were used to complete the initial fabrication process. The "Z" sections were placed close together and any gaps between were closed with good-ole duct tape. With the body still inverted the "Z" sections were almost horizontal and lay-up of the laminate was very easy again layering in several layers of glass mat and wetting out with resin. The legs of the "Z" forms were approximately ½" wide allowing for the riveting and not having too much area to cover with glass to the parent fender material. Note this encapsulates the rivets for added grip and strength. Righting the body presents a pocket again almost horizontal to fill in with mat and resin after removing the duct tape and grinding off the rivet heads. The contour of the filler between the two edges could be shaped to a complex curve, but I opted to keep a more or less flat profile, KISS. Besides this was my original intent from day one.

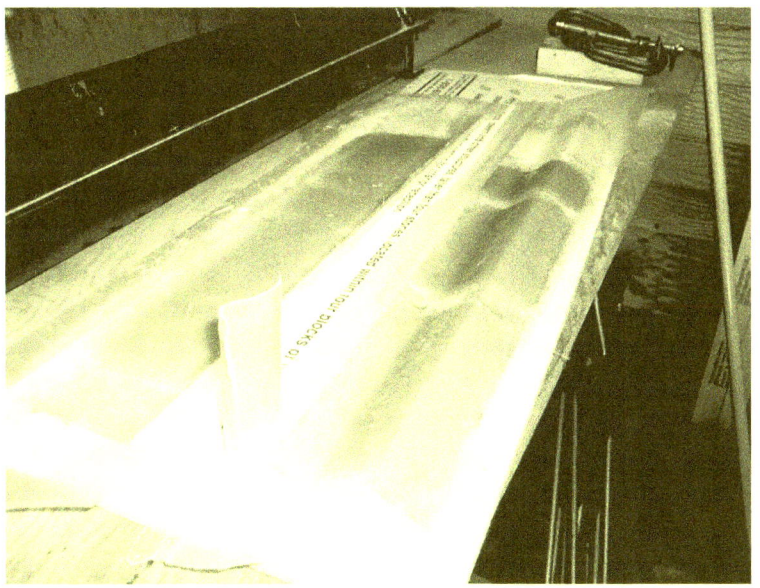

Fabrication of the "Z" shapes for mounting/closure of gap.

"Z" shapes installed with pop-rivets, ready for fiberglass mat. Note all surfaces cleaned to glass base material.

Top side ready for top laminations of fiberglass.

Underside of rear quarter panel shows spacing and cut line.
Flat sections of fiberglass pop-riveted in ready for glass
(mat and resin) inside and out.

In keeping with the idea that modifying the Europa S2 into a convertible or del Sol, in my case, body strength and flex is always brought into play. As covered in a previous segment the roll over bar and pillar improvements were implemented to add strength to the body. Also mentioned in previous notes was the presence of rust, including the bracket shown. This bracket is the side seat belt attachment point inside the lower door sill. The replacement bracket was fabricated as shown using left over 1" square tubing and ½" nut (and 1/8" plate) welded to a formed 16 gauge steel plate. Standard installation has this bracket bolted to the lower fiberglass cockpit tub sidewall inside the lower door sill.

The outer body shell and floor tub are riveted together at a lip below this bracket. When planning to cut the roof out I opted to add strength to the body structure by installing a 1/8" thick by 4" high plate of aluminum the full length of the body between the wheel wells. The aluminum plate was sandwiched in between the fiberglass and seat belt bracket.  Finally to integrate the structure for strength the roll over bar was also bolted to the fiberglass where the typical upper diagonal seat belt post is located and to the new outside seat belt bracket.  There should be no flex in this body when completed.

Also to beef up the fiberglass floor panels behind the driver and passenger seats steel sheets were fiber-glassed into the floor under the twin 7 gallon fuel tanks. The right side of the car had some damage to this panel and the left side was severely damaged on the bottom as this is the normal area for removal of the fuel tank on the S2. Since my fuel tanks will come in from access holes in the fire wall the bottoms can be fully enclosed.

Seat belt brackets original and much stronger replacement. 1" square tubing replaces 3 individual mounting nuts also bolts onto the roll over bar for continuity.

Left side fuel tank floor broken fiberglass shell.

Replacement fiberglass sheet; ready to install.

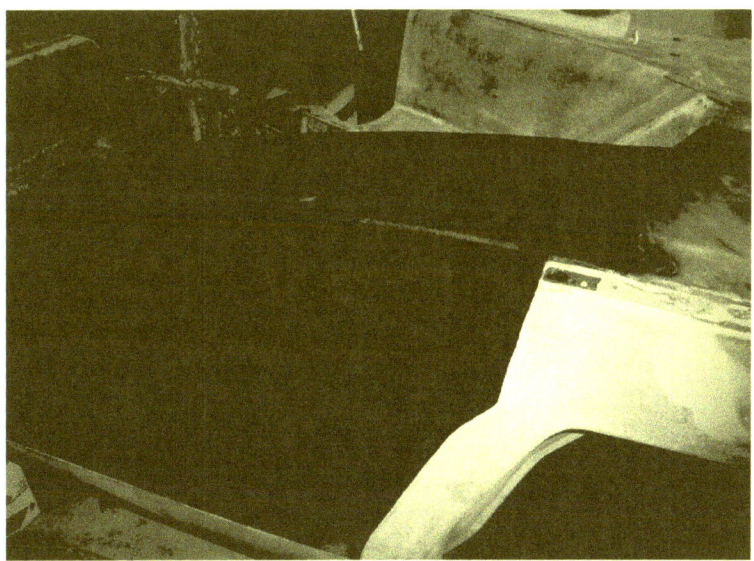

Completed closure and repairs to left side fuel tank body shell. Note these repairs were done before the rear fenders.

Xxxx" "The night was ~~torrid moist ...~~ XXXXX
I finally remember where this line came from and speaking
of lines. Several of the body lines on the Europa S2 were
broken and needed repair.

Left front fender flare was damaged. One Lotus
expert presumed the damage occurred because someone
tried to lift the body off the chassis without removing the
crucial six (3 per side) bolts holding the body to the
backbone chassis in the cockpit. Damage can be seen in the
photo, not bad enough to require a new flare but bad enough
to require some mold to hold the form.

Two methods are common, one is to pull a mold
from another undamaged car or employ some formable item
such as clay or in my case the continued use of nylon rope.
Clean up the wound to a beveled edge with clean glass
inside and out. Shape a piece of rope to fit the opening and
check the shape using a mirror image cardboard cutout of
the other side of the car. Glue and set the shape with resin,
shaping the remainder of hole with old reliable, "Duct
Tape", then apply the required two to three layers of
fiberglass mat, laminate to the original fender and sand and
fill any remaining holes. I worked only on the top surface,
leaving the underside and back for later when the body is
inverted making for an easier lay-up.  After feathering in the
glass, spraying with primer helps to find any remaining pin
holes to fill with glaze.

Once the body is turned over the lower lip of the
fender flare is easy to layer in with glass. At this time the
rope, (or other form) can be removed or in my case I
encapsulate the rope for added strength to the repaired area.

A similar method was used to remove the "sneer" of the inlet nose cone. When last fixed the repairer did not make the two sides symmetrical leaving the air inlet looking strange. Again a mold is configured, this time I used heavy cardboard shaped to the good side of the inlet and reversed it to form a pattern for the bad side. The resulting "U" shaped piece of fiberglass was molded to the pattern then inserted into the opening and filled in to correct the "sneer". While the body was inverted it was noted many other areas of broken or irregular fiberglass sections needed repair. The entire bottom of the front "Boot" was broken so after a cleanup a layer of glass mat covered the entire front and resin was poured on to smooth the surface and mend cracks. Also as can be seen the footwell holes for pedal (brake) was broken out, this due to my inability to remove the transfer pivot box of the brake assembly. Due to rusting the bolts would not budge so I ripped it out needing to fix the glass later. Areas around the radiator needed fixing, at this time I squared the opening and will later install a closure plate because the radiator is being moved to the rear of the car, Get those hot radiator pipes out of the center counsel.

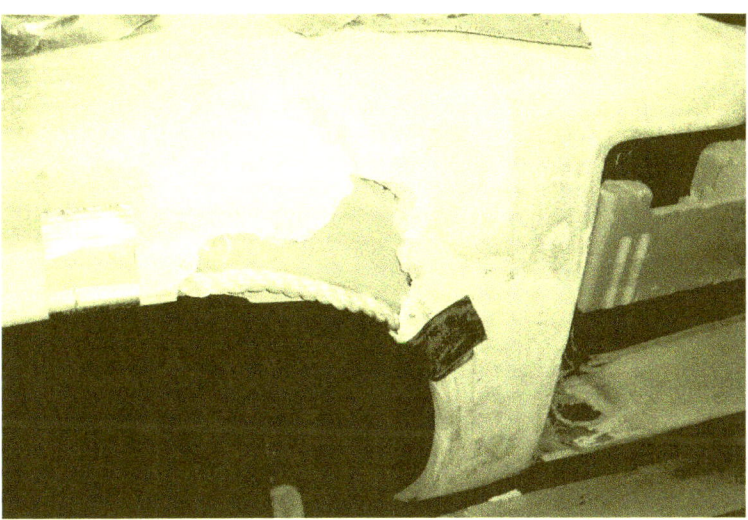

Rope on fender flare backed with Duct Tape (Racer Tape)

Glass repair to flare; primer to find flaws and repair holes. Note bottom lip has yet to be added.

Bottom of front boot with layup of glass mat and resin.

The answer to the riddle is: the comedy classic movie "Throw Momma from the Train". Billy Crystal a writer: could not get past this opening sentence due to writers block. Old age affects memory and becomes CRS in this writer too.

Xxxx" "The night was ~~torrid moist ...~~ XXXXX

Paul Quiniff of Fiberglass Solutions in working garb, item under his ministrations is del Sol's dash board.

       Things just seem to hit me funny at times and I go off on a tangent. This particular chapter came from seeing what could I, do with the deteriorated dash board of the Series 2 Europa. Not much it seems as total disintegration has occurred. Purchase another unit this one from Phil another club member, with gauges wiring and other dashboard appendages. Even this dash was in only fair shape with the usual cracking and dislocation of lettering, plus this is a dash from a much later Europa, a Twin Cam Federal. Key giveaway was the four small gauges for Fuel, Amps, Oil and Water are not together in the center of dash, the heater controls separate the gauges into two groups of two. This presented one major problem; the mounting screw location. One could relocate the gauges to the center or my option to relocate the two inside mounting bobbins. An angle iron bracket was fastened to the outer holes and inside holes located on the bracket. The bobbins on the fascia board for the two inner screws were cut from the body and relocated to the angle iron bracket and again fiber-glassed to the fascia board. Now the update dashboard will be properly located.

The recessed heavy duty box board gauge cluster and glove box are not to my liking so they were removed and these openings and radio slot were plugged with plywood. Hot glued to the dash and filled with body putty sanded smooth before re-drilling the holes for Tach and Speedo. A final step in making a Flat Dash was to add side wing pieces and filling in the dip in the center top of the dash, again with plywood. Added width to dash to close dash to body, stock closure is by crash pad. Final step will be to refinish the dash, but how and what kind of finish would do. Fiberglass or paint might work or even some type of veneer. Would this be classy enough or special, Heck NO!

What really would stand out would be Carbon Fibre. Black and shinny with an awesome pattern. I put the idea to Paul and asked for a guesstimate of price. I could live with the $$$$$ and would really like the finish. The procedure to get a good finish demanded that I not do the work. Let an expert, called Paul, handle this one; albeit I did do a lot of the prep and un-glamorous work. A "clean" flat steel layout table with no bumps or bruises, covered with a sheet of heavy duty plastic, and coated with a parting wax form the base for the lamination. Followed by two layers of carbon fiber mat, two layers of fiberglass mat, and copious amounts of resin; wetting out to smooth the multi-layered sandwich. The face side of the dash is located on top of the composite and weighted down with heavy steel bars to keep the whole assembly flat and evenly compressed. Believe me Paul works with the actions of an Artist insuring everything is just right.

After a night of thermal set up of the resin; the edges must be trimmed and all the holes opened up. Again Paul had all the right tools (hole saws & grinders) and technique to prevent cracking which would ruin the finish. Drilling from the back side to locate the holes and only going part way through flip over and complete drilling out the gauge and other holes. I took my dash home and immediately installed in the car after sealing all the edges to keep any moisture from sneaking into the laminate to prevent de-lamination; which can occur. The main reason to install in the car body was to ogle the appearance and secondly to keep away from any disaster in the garage while waiting for gauges and wiring step to be completed.

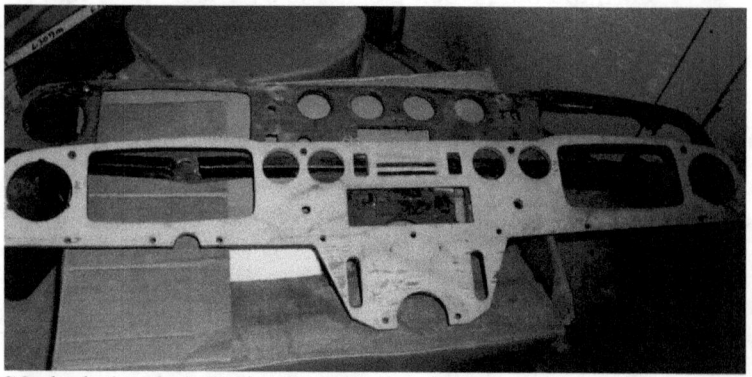

S2 dash (under) Twin Cam Federal (top). Stripped of gauges.

Dash back side, note: plywood panels at glovebox, radio and guage cluster. Black area is carbon fibre before trimming

I'm saving photos of the finished project for later, to tempt readers to continue with the saga of Europa del Sol.

Xxxx" "The night was ~~damp moonlit~~ XXXXX

Why did I buy that bench top media blast booth? It could not have been for the alloy wheels which are #1 on my to bead list. Even at 13 x 5.5 they are TOO BIG to fit inside the doors. And I have done more than my share of modifications on this project to start changing the machinery too. I guess some of the smaller parts; the pedal assembly and suspension parts, done recently, count for some usage. So off to see the wizard, from another movie, OZ: for a larger machine. While Paul poured out some of his famous resin I used the booth to blast 35 years plus of crud from the spider factory alloy wheels. Remarkably the wheels are in pretty good shape and were it not for aesthetics we could stop right here. Again Paul came up with a good tool for me to complete the project, a small fractional horsepower gear-head motor set up with a hub assembly were enough to mount and slowly turn the wheels to sand and polish the rim and webs. Once polished to a brite finish each wheel was sprayed with clear coat several coats inside and out, front and back. And to conserve space in the garage the tires were mounted and balanced to the fresh rims. Typically the wheels have a black accent finish on the inside but "KISS" not having another step plus a softer appearing wheel which has a gray cast against the polished rim is accent enough for current planned car exterior of a blue-grey color, open to change at drop of a hat.

So here is the status going into start of second year for del Sol. The almost complete body shell is on the carport with a fancy carbon fiber dashboard and no other components mounted. Preliminary gray primer and red polka dots from glaze filler giving a "chicken pox" appearance over the entire body. Doors, bonnets and glass stored up above the garage rafters, with other sundries still in rough shape needing surface preparation. The Gordini motor partially disassembled with its components scattered about the shelving of the shed. The wiring loom, on a work table is in partial disarray being a bastardized version of the Federal Special and my idea of a S1-. Many wires removed as not needed; power windows, pollution control, seat belt warnings, etc and others were added or rerouted such as cooling fan and temperature sender to rear of car.

The two frames are in the process of restoration inside the garage with all but a few items removed from the original frame, Renault motor and trans on the side as well as all the suspension components. All these parts came off with the usual grunt, liquid wrench, and heat where required. Only a few licks with hammer and some application of the saw were needed to free up shocks and bushings. Steel has been cut for seat frames, some chassis replacement and restructuring (make stronger and free of rust). Both frames need work on the front "T" section a notorious problem area with Europa(s). They really do rust out and have ample pockets to collect and trap water: plus the metal is a thin gauge from the start.

Another piece of the puzzle in the works is the interior. Finding several nice skins at an upholstery auction the fabrication of new seat cushions and seat sides is being done inside the house where it is nice and warm 74 degrees versus 36 degrees in the garage. Who can blame me?

One other useful tool for this project has been the portable stand or roller cart to move the body shell around. Fabricated from 2 X 4's with cross bracing and a set of 3-4" wheels 2 fixed and 2 swivel allows movement in and out for clean up, working in or out of garage. The entire frame is bolted together so it can be easily stored when not in use.

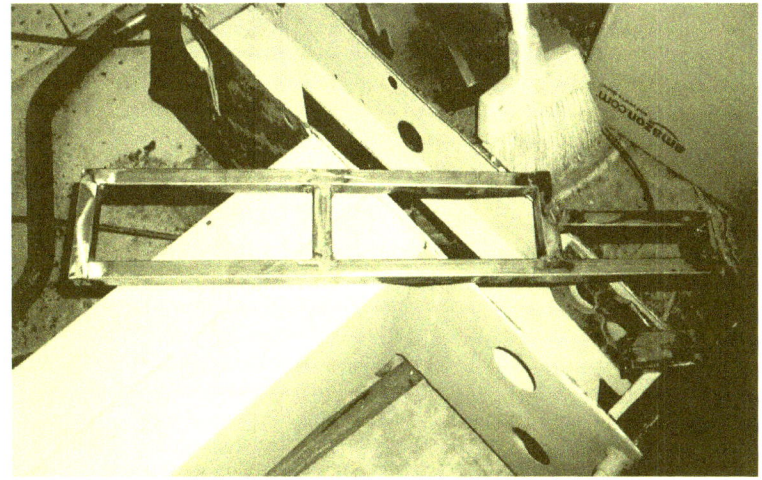

Square tubing weldment ready to install on original S2 frame. Notch in tubing is for installing brake mounting plate. Grey color of frame is rust stop primer.

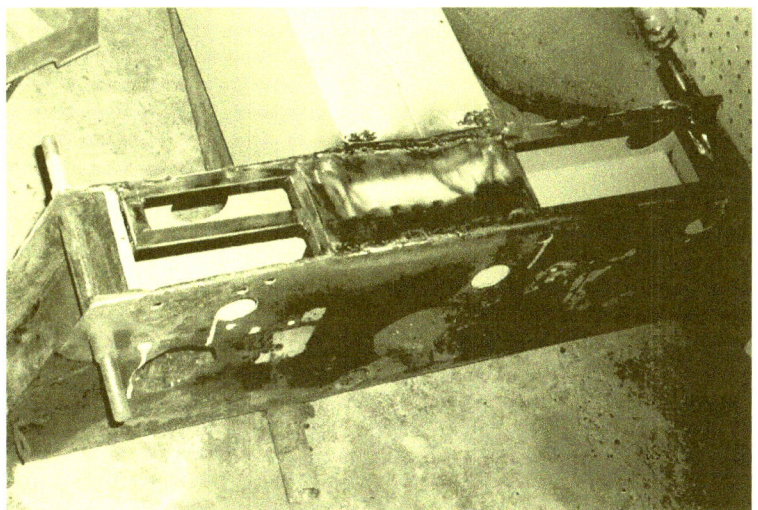

Bottom view of new square tube section welded to original frame to add strength. Later pieces will enclose and box in.

Although already covered this segment is added to stress the importance and effort in mending the chassis.

The backbone chassis of the Europa is important and as so I find it more important to fully cover all of the base work done on the base. One basic premise of any Lotus is light weight while maintaining a strong stiff chassis. To allow for a minimum amount of steel ergo weight thin gauge material is used. In engineering terms structure shape or form determines the load carrying characteristics of a frame. The sphere is the strongest, followed by a circular shape then comes the box or rectangular form. The backbone chassis of the Europa and other Loti is a thin metal rectangular shape and as long as the four sides are solid and connected it remains strong. The amount of rust and degradation of a typical Europa chassis not cared for was amplified on del Sol. The forward "T" section carries the steering gear, front wheel suspension, brakes, and other components like brake lines, cooling tubes and heater lines.

The fiberglass Europa lower body forms a 3 sided shape sliding over the rectangular form and enclosed on the 4th side with closure plates. Unlike other car frames of heavy metal the Lotus counts on form over weight. Problem is some of the form pieces are traps for moisture, leading to rust. To strengthen and repair the frame; new heavier gauge steel was welded in at problem areas and where totally wasted; new required gussets and heavy wall tubing were installed to weld and or rivet new sheets in place.

Above photo shows one such spot where water resides and rusts through. Fresh steel is welded in at all such spaces. This location is at back of cockpit ahead of the engine/powertrain.

Other small sections include the two side motor mounts already discussed and the rear motor/transaxle mount.

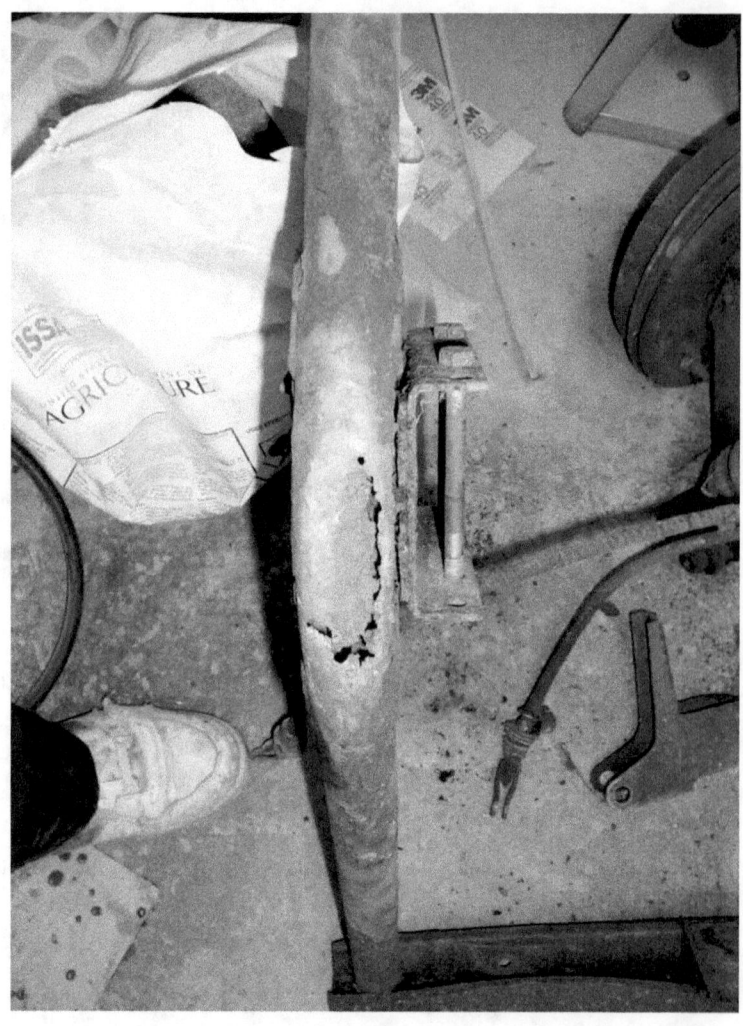

     The worst location requiring attention was the front "T" section. A framework of 1" square steel tubing was formed for the bottom and top welded in place at the corner joining the top, bottom and sides allowing for heavy metal to be welded and riveted to the original rusted and new laminated sheets.

Bottom of "T" square tube frame welded to parent frame.
One each top and bottom to beef up edges.

Front top of "T" section ready to receive added steel.

Steel sheet riveted to top section later welded to side sheets.

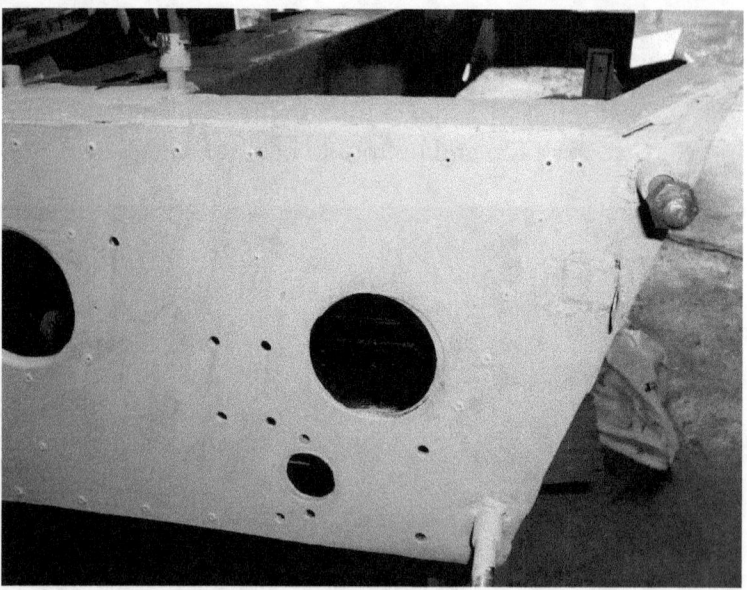

Front sheet steel riveted and welded to match all holes for mounting brake master cylinder and steering rack.

Pocket form on each side is mounting pickup points for
"A" arms, this area is strengthened further.

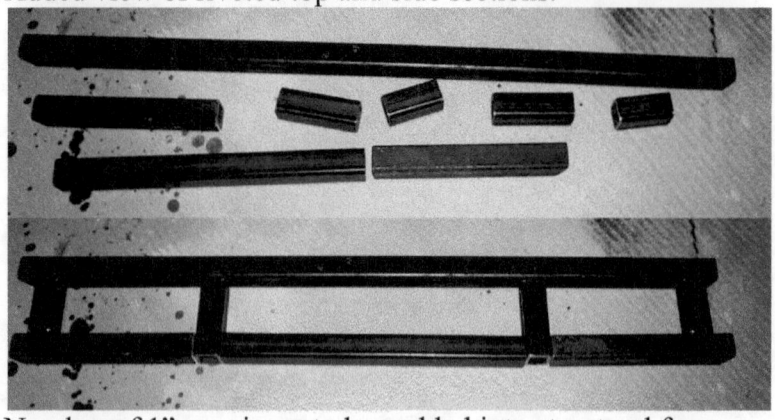

Added view of riveted top and side sections.

Number of 1" sq. pieces to be welded into structural form.

At "Y" section of frame added steel plates correct rusted bottom section and bulkhead plate for heater pipes, E-brake cables, gearshift tube and other feed-through repaired.

Layout of frame at rear is a "Y" section where engine and transaxle reside.

Xxxx" "The night was ~~night~~ XXXX

Some of the old adages and sage advice are often forgotten. This is especially true in my last adventure with del Sol. Both of my Grandfathers were German craftsmen true to the old tradition of the homeland. One a tailor and the other a carpenter both would tell me "Willy", that's what they called me, "Measure twice and cut once" and of course the old tailor story of "I cut the pants down twice and they're still too short". And definitely in my case the old saw, "We grow too zoon olde und too late schmart", poetic license from the German.

These all applied when working on my seat frames. Many months before I started to do the physical work on the seats I carefully laid out and drew to scale all the pieces required to duplicate the Lotus framework. The steel stock was cut and stored for later use. During the body reconstruction I needed some spacers and fortunately the flat stock which was to form the top of the seat frames was just the right size and luckily we had cut several extras knowing this was a tough bend scenario, (make that impossible) for my bench top bender. Well I chucked the drawing and started on the seat backs, took a couple of quick measurements and started to cut, OOPS, when I finished stripping the fabric from the top a big mistake came up. I had taken a straight line from the center bend to the top frame when it is actually a two bend, a double apex in racers terms. So I shorted myself of some width in the mid section. I can cut all new steel, weld on another piece, or better yet graft on the top piece from the original seats. The final one is the best solution since: #1 both top frames are in pretty good shape no rust and all, #2 forming these channels will prove difficult, and #3 both the frame backs have about a 5" diameter hole formed in a rolled edge of ¼" flair or nacelette with a primary function of stiffening the sheet metal back like adding a cross brace, something called "section modulus" in engineering terms. Bet you thought they were for stereo speakers. Forming a rolled edge like these is difficult so grafting to the seat tops was a good solution to my CRS or old-shimers event preceding the need. Good Save, Bill!

Another faux pas was the middle seat cushion which I made square and in reality is tapered, Try Again. Because I used all of the original leather as a template for the new, the hide is cut right but need to cut a new backer board and foam. Found this when the gathering and stretching the leather did not go well, too many wrinkles. Hopefully when all the parts join; steel, fabric and foam these mistakes will all be covered up or corrected. Maybe I needed a grandparent who was an upholsterer.

When reading the body of text how many of you read between the lines for good digs at the author? I'll add them now. "Bill you have plenty of width in your mid-section" and we've seen you drive, "You always miss the apex". "Those wrinkles are looking back from your mirror". "Short is such a little word". Or better yet, "Your grandparents should have practiced birth control". Happy Now?

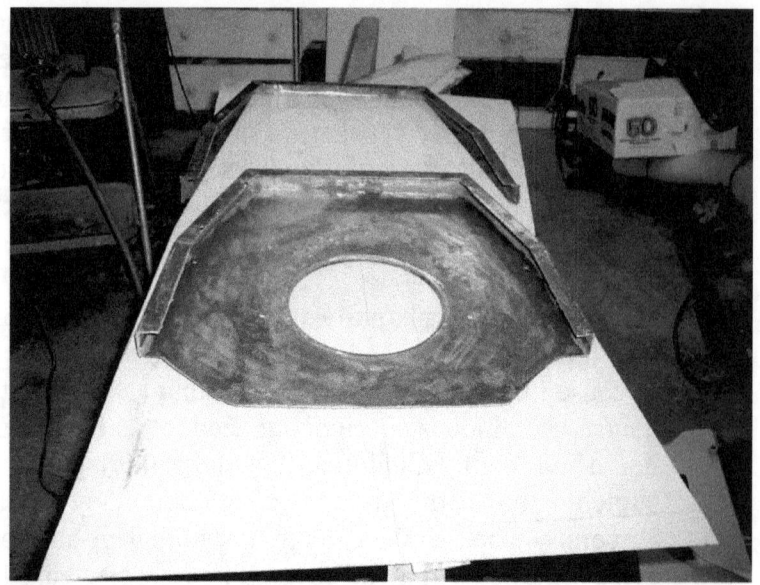

Top seat frame saved from original seats to be grafted to new metal. Note difference of shape and space.

New material appears white in photos, this is coating to prevent rust and make paint adhere better than naked steel. New metal is short at the apex.

Old vs. New. Rust replaced by new steel, sans bottom.

New bottom; ready to weld into side frame.

Xxxx" "The night was ~~not a normal night~~ XXXX

The #$*&^ notorious door hinges. What can a
Europa owner say about the door hinges but curse a lot
aloud or otherwise.  Del Sol was no exception but did, as
mentioned previously, need to be removed by hack-sawing
the pins from the top. No problem on the bottom as the
bobbins had broken out on both sides.

Extracting the bobbins from the bottom end of the
hinge rod took much convincing, hammer and tong, then
cutting the surplus fiberglass from the bobbin before
reusing.  After procuring a set of stainless steel rods of the
½"diameter persuasion, I fit the rods into the still in place
top bobbins and inserted the bottom bobbins to examine the
fit and to figure a method of re-glassing into the body shell.
Clearing away broken/damaged fiberglass from the areas,
left and right sides, a cardboard stage was built to support
the bobbin in the proper orientation. The new rod was a
snug fit and would locate the bobbins while the glass cured,
but, wait! What if the glass/resin grabs hold of the rod, after
curing, in particular inside the door sill, how to extract? Cut
off and drill out a stainless steel rod (not in this life), break
the bobbin from the glass again (never), who knows what
could occur. Without the rod for alignment it could prove an
even bigger problem. My solution was to replace the
stainless steel rod with a nice straight piece of wooden
dowel rod. Should this dowel get trapped no problem
drilling out wood.

A generous amount of fiberglass mat/cloth was cut
to fit the opening to encompass the thickness of the bobbin
and then some, top and bottom. Wet up as usual and
perform a typical layup/lamination.  Once cured the wooden
dowels came out without a hitch and the excess glass from

the top was removed, ground down to just level at the bobbin. The cardboard stage was removed and the stainless rods were again tested for fit. Volia! Perfection! Next step is to decide what form the keepers, cotter pins, abrasive washers, and hardware will take when the doors are replaced, original, Stainless steel upgrade or something other? This; my friend is for another day. Now, it's time for a cold one.

P.S.    A trick learned a long time ago when "Orange Crate" my brilliant Mack Truck Orange, '73' Twin Cam Europa, was repainted by Elf Specialties. Ed Fortin cut out the top of each fender above the door post, four inches square, to gain access without removing the doors. Repaired the damage and cupped the top bobbin fiberglass to hold lubricating oil. Also he drilled out an access hole above the top bobbin, inside the door sill, closing with a plastic hole plug so new oil could be added later. This feature and the CB antenna stub helped me identify my old car when it showed up repainted red at a club function years later, in the possession of our Corps member Justin E. A well traveled and repainted car, original Lotus yellow from White Bear Lake, Minnesota by way of Joliet, my Lotus yellow with red, white and blue British flag on front bonnet with matching roof and rear deck racing stripes, Mack Truck Orange and finally Red. Who knows where and what color it is today. I need to check again on the internet Europa Registry it may re-appear, Purple or Black, in West Overshoe, Texas?

Picture shows the broken out bobbin still attached to door once top hinge rod was cut from opening. This is part of reason Lotus doors sag and misalign. It is not interesting to photograph the repaired section just a nice smooth hole.

Xxxx" "The night was ~~Friday the 13th~~ XXXX

Science is a wonderful thing, too bad some people who work on motorcars are not too wonderful. Science is responsible for giving us tools to work with and depending upon the tool and the operator of same; the results can be very different.

Take the del Sol frame, nature provided it with furry creatures to crawl around the inside of the frame depositing acorns and nesting materials. Not a big problem, just remove the debris and move on. Nature also provides the process of oxidation, better known to steel as 'RUST'. This is not as easy to move on from; to a better world. One high stress point of the Europa frame is the engine mount attach point to the frame. A large amount of rust gathers behind the reinforcement plate which is spot welded to the outside of the frame. Rust has a habit of expanding and layering to buckle and distort the shape of metal. At exactly the same time it eats through the thin metal of the Lotus framework, go figure. At any rate fixes depend upon location and extent of damage which in this case was extensive enough to require cutting out a section of the frame behind the reinforcing plate. A new 14 gauge steel piece was grafted in, continuously welded on both sides and ground smooth at the engine mount so as to not change any dimension. Now back to our science lesson. Tools used included: a 4" grinder with cutting blade (electric motor, gears, bearings, and carborundum composite wheel); saber saw (same motor as above plus tool steel blade); 'C'clamps (screw drive which in physics is simple mechanical  device – the incline plane wrapped around a hub). Final tool the welder is a bit more complex, a power supply AC/DC, drive motor for wire feed, the alloy wire and shielding gas Argon/$CO_2$ mixture. Whew too much science; Physics, Electronics,

Chemistry, Metallurgy, Mechanics. All I wanted to do was work on cars with an Internal Combustion Engine (not motor). Motors in the true scientific terms get their power from an external source like electricity. Only electric cars have drive motors to motivate them. Most Loti have: fan motors, wiper motors and window lift motors but no Gas Motors. School lesson over, except for Psychology, for the nut who tried to turn the center (rear wall of the frame) where shift tube, cooling pipes, and e-brake, clutch and accelerator cables; frame into a Picasso art sculpture. In order to straighten this piece I had to attack the frame with the reciprocating saw. Remove the offending segment, flatten it and weld back in place.

By the way whose brilliant idea was it to weld the heater pipes onto the frame, maybe using the steel pipe for strength?

Another application of the incline plane is seen in the photo. The standard frame is easily tweaked by incorrect jacking, running over or into objects, or lame mechanics trying to force something into position. At any rate the use of clamps with heavy steel bars, lacking a hydraulic press/clamp, can be used to convince the steel back into the intended shape. The Lotus frame can easily be harmed by above mentioned machinations and result in poor alignment of wheels, etc. which could affect handling. Remember this next time you run over something or somebody.

"Alls Well That Ends Well" (that's Literature) and Frame #1 is done, now moved on to original del Sol frame described above.

Frame #1 pictured below was also de-rusted, welded at a few cracked spots and had reinforcement done to entire front "T" section. 14 and 16 gauge steel sheets were fabricated and laminated to existing rust pitted and weakened metal. All primed, riveted and welded to add strength to front end of this frame. Adding a small amount of weight cannot hurt performance too much.

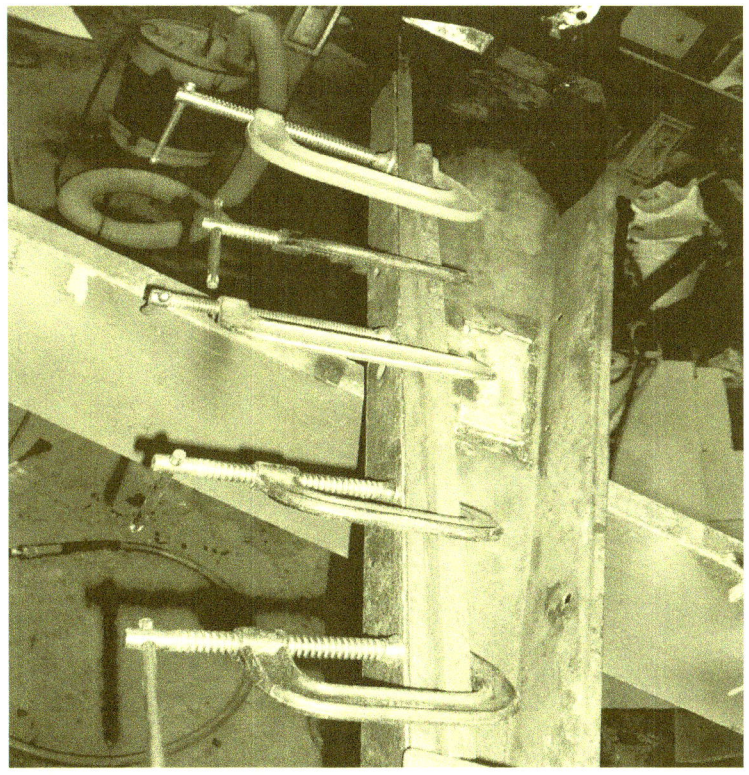

C clamps and bars to straighten frame by patch welded to side of frame at engine mount. (del Sol frame) Note a comment by Steve Styers on seeing the above "You can never have enough clamps"! How terribly true.

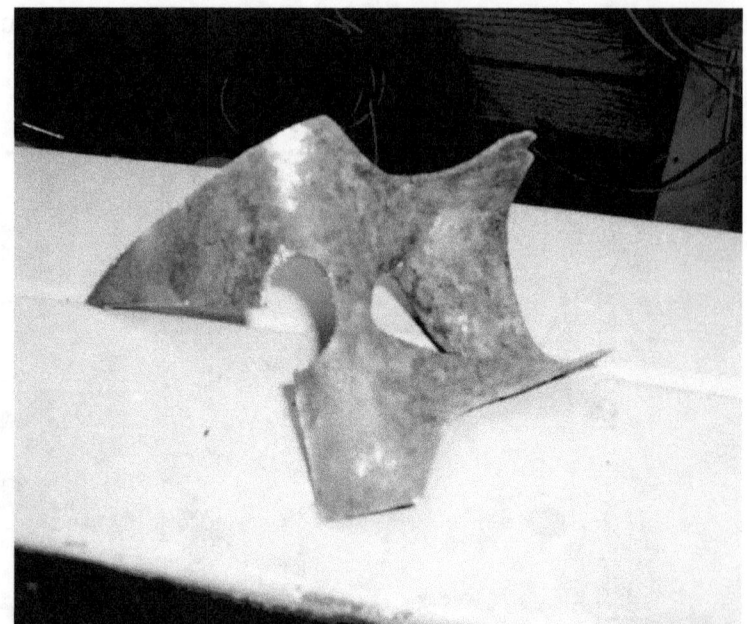

This is not a sculpture from Daley Plaza; this is act of a madman with hammer and cold chisel. This is shape of part removed from del Sol frame.

Spare frame, reinforced, redesigned, rust stop primer, red paint and ready to go. Where? Races! Ralleying! Riding! Could these be the three R's of Lotus?

Xxx "The night was ~~later than Friday the 13th~~ Xxx

Lotus gearbox-transaxle. Clockwise or counter-clockwise, which way does the engine turn, could never remember, but once installed the remarkable Loti had four reverse gears and one forward, fine for super stunt driving. And the car apparently thought it was a front wheel drive car with the engine up front. Re-assembled transaxle found final drive gear on the wrong side of the output shaft. In actuality this is one reason the drive train of the Renault was chosen; a simple switch of the ring gear reverses the drive allowing the R16 front engine moved to the mid-engine aft mounting of the Europa. The re-assembled transaxle had been put together backwards. My gaff here required the complete removal of the drive train. One purported expert suggested to just split the tranny halves swap the gear and put two sides back together. No can do; as several bolts at bell housing holding clutch and transmission in place enter from the clutch side of the bell housing. So need to pull the whole assembly leaving the clutch and flywheel in situ. Needless to say this was not the last occasion of removing the whole ass end.

After re-re-building and re-re-assembly the chassis was converted to a test bed for the previously rebuilt R17 Gordini Hemi-Head. During the test session a small oil leak reared its ugly head or butt. Nothing unusual for a Europa Twin Cam, not so with the Renault. The leak into the bell housing would potentially trash the new clutch. Soooo call out the transaxle pit crew; over the wall, getting good at pulling the stub axles, remove starter, and remove cooling system and this time to access the offending rear seal; remove clutch and flywheel.

It would be ingenious if a zipper could be installed so quick change could occur. Only saving grace is the body has yet to be mated to the chassis giving very easy access to most components.

Next fly in the ointment; the supposedly non-re-buildable, (old wives tale), water pump began dribbling coolant. Not a gigantic leak of coolant in leakage proportion but required dis-assembly of the cooling system; Again! The VW Jetta radiator gracing the rear placement in a fabricated aluminum angle frame; sits astride the original Lotus frame, transaxle and exhaust muffler. Internet search found a rebuilder of Renault Gordini R17TS water pump back in a few weeks. This new flaw put a halt to power plant testing. Time to back down; look at my options through my cooling system, a glass of sudsy amber liquid, gift of the gods.

Second rebuild transaxle on floor, flywheel on box with new yellow markings at timing marks, TDC and 8 degrees.

New rear oil seal in, ready for assembly.

Renault Gordini test bed, easy access before body comes.

Xxx "The night was ~~later than Friday the 13th~~ Xxx

Engine rebuild and rework. The original R16 80-90 HP S1-S2 Europa was a fine choice as a starting point for the Type 46 Lotus for Europe. The cars' diminutive dimensions and light weight became an adequate marriage. The 47 or race version given a GT designation grafted in a Rover V8 and Hewland transaxle, too much for a street machine. Later versions of the Europa Special and Twin Cam adopted a Kent Ford bottom end and Lotus twin cam aluminum head generating over 100 HP and easily modifiable version of the Formula Ford engine.

An alternative is to upgrade the R16 with the Renault R17TS High-Performance Hemi-Head. Fortunately one of this breed not too tired was purchased from another club member. This Canadian version of the Hemi with a modified fuel system from the standard fuel injection to dual Weber DCOE 40's to allow for optimized fuel control and also look so cooool, defining race, sport engines of the 60's, 70's and 80's

The base Renault engine is an aluminum block with wet liners of cast iron. Only minor problem found on the existing engine was cracks on two of the liners at the lower end, below the piston/rings. The liners were welded and true bored and returned to service.

Keeping with my original commitment to use where practical all existing parts; all existing bearings, shafts, pistons, rods and internal engine components were recycled. The 807-13 should produce 120-130 HP when done. The top end of the engine was refurbished; cleaning up the intake and exhaust valves and ports. Regrind and lap valves and reuse the rest of the valve train.

On assembly all gaskets were replaced with new as is normal practice focusing on clearances of the wet sleeves with respective paper spacer ring gaskets in the block, an oddity of the Renault engines.

Critical part of any OHC/OHV engines is the timing chain. The chain has a fixed length without a connecting link. The chain and sprockets were worn and had to come out, the crankshaft sprocket would not leave home.  The chain and sprocket needed to be cut off as no amount of heat and/or brute force would budge.

The external components mounted to the engine presented several problems and questions of how did this engine ever perform. Dual intake manifolds joining the Webers to the R17 were not uniform. Alignment of the throttle shafts was impossible resulting in another trip to the flannel shirt club meeting engaging the milling machine to true up the two manifold components, now within a few thousands acceptability.

The original R16 configuration is intake and exhaust piggy backed on the driver side. The Gordini is a true cross-flow head so the existing exhaust system does not adapt. The ½" thick manifold base plate and 4 oval ports outlet to round header pipe transition pieces came with engine. Past experience with Kermit my 67 Series 1 header pipe ran down and close to the starter, cooking the starter solenoid. Therefore the custom headers made for del Sol go up and over the engine then down to a second hand, ebay purchased, almost new stainless steel Stebro exhaust muffler system. The Stebro mounts to the transaxle and bell housing exiting through the center of the rear body grill opening. Question of clearance and being able to install or remove the muffler through the opening was later proved to be doable. The special custom made headers were expensive but gave further formula car appearance. Again very good look.

Weber DCOE-40 carburetor intake manifold tubes dimensions being corrected to true on milling machine.

Chassis on display at BritishCarFest, Note the headers, rear mounted radiator and Webers of Gordini powerplant.

So many parts of the body needed to be re-glassed a special section is devoted in addition to others covering the many optional upgrades, i.e. del Sol mods.

Both deck and hood lids needed a fair amount of work. The front hood had large chunk of Bondo at the lead edge and some of the sides creating misfits. Also the hood had a ton of fiberglass crazing, cracking. Once the layers of paint and fillers were removed it was decided to a layer of the lightest weight of mat and resin cover the entire hood surface. The rear deck lid did not need a full layer.

Once both surfaces were free of crazing they were fitted to the body to check for fit. The front hood had many areas where the fit line between the hood and cowl or fenders was off. Each segment was addressed separately building with new glass mat until close then layers of Bondo. Worst problem was the very lead edge and at least 6" back was the height of the surface being at least 1/4-3/8 inches low. This section was built up to level in steps.

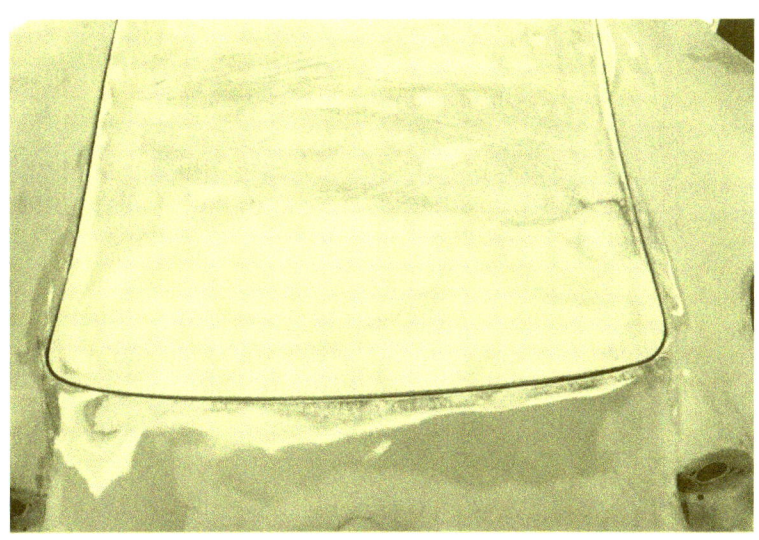

One problem with all that added glass was weight so to reduce somewhat a series of lightening holes 1" diameter were drilled in the side and rear lip of the hood.

Hood at paint shop shows the lightening holes to reduce the overall weight of the segment.

Biggest problem with the rear deck lid was what appeared to be a warpage that required saw-slotting the lower lip of the lid set in a fixture to straighten and re-glass the entire length to lock into straight line.

More fixes include several broken segments of the front foot-wells where removal of body caused these because the pedals and bell crank would not be freed.

Photos on next page show the item of fiberglass repair to the bottom of the body. Extensive repair to breakage requires cleaning both interior and exterior grind down to base fiberglass and clamp into position true and square.

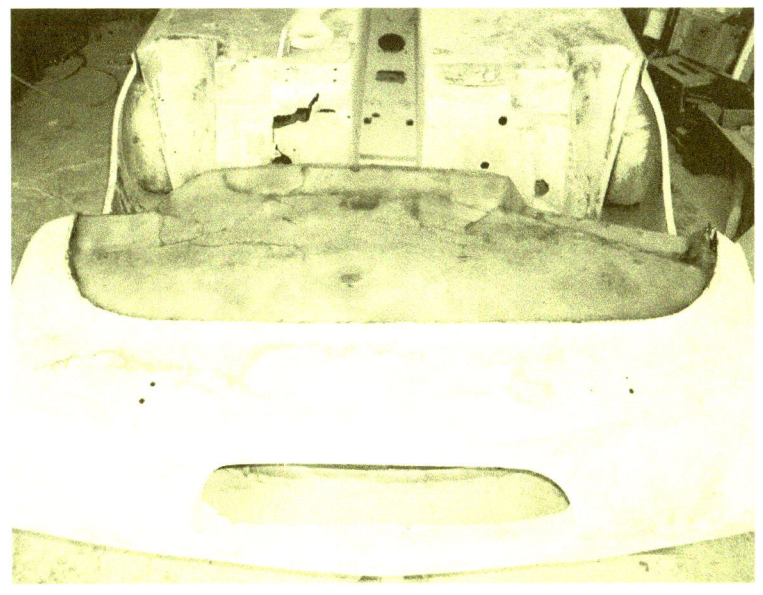

Lots of repair needed to eliminate breaks.

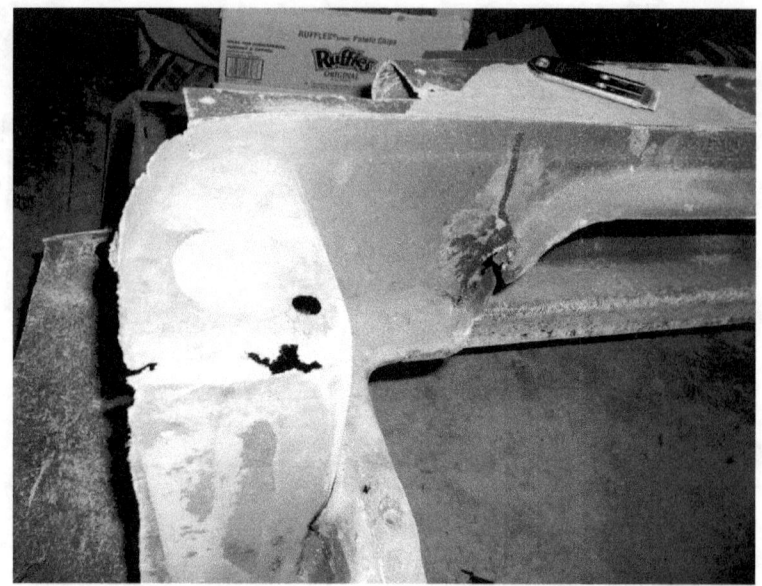

Rear wheel well and rear cowl/grill opening needing repair.

Original radiator opening in right wheel well was filled in completely as radiator cooling system was moved to the rear hatch. Eliminated the cooling pipes; which added heat to the cockpit; running to and from the engine bay.

Ah! The moment of truth OR did whatever changes to either the body or chassis make it impossible to reunite the two major parts of del Sol. Can we again arrive at a cohesive machine? Call to action or "Mini-Clinic" a catch phrase or "codeword" for lets all get together, as many club members as possible, sort out a few problems, learn some new techniques, offer assistance (man power), ….Bah who am I kidding we congregate, BS, drink way too much male bonding fluid and fumble into car solutions.

Anywho; the gang assembled to assemble. Using the engine hoist as an ersatz cherry picker we lowered the body, complete but unpainted, over the chassis with the usual squeaks' and squeals of fiberglass grudgingly giving way. A few offending areas were found. Namely the Datsun B210 master cylinder mounted to the bell crank angled brake pedal had to come off totally. Various holes that were re-glassed due to previous fractures were again located for steering column and brake pedal and re-drill several body mounting holes. Parking/e-brake crank arm; hole re-done.

Proved that full chassis assembly needs to break down and remove components for later re-install once the body is in place. Items are: swirl pot, Webers, muffler, e-brake crank arm, master cylinder and brake pedal bell crank housing. Also proved all the modifications with rear radiator, carbs, and engine all fit in an acceptable manner.

Notes were made on items needing work with body off before paint, then the body was removed back to the wheeled assembly cart. A few hours well spent everything tucked back into the garage for another day.

Swirl pot must come out, while body and guys hang around.

Alfred and John try to talk the body down,,,slowwwly.

Body fully down, add boot cover, check clearances and latch hold downs. Dennis practices sign language to shutterbug wife, Nancy.

In preparation to receive the body much needs to be completed to the chassis with systems that will be covered once the body is on: almost inaccessible.

One such component is the braking system. The master cylinder dual chamber was examined and found wanting. The same mounting and piston/reservoir are found on the Datsun B-210 master, new and reasonable versus a rebuild on the old Lotus part. The remainder of the brake system lines, fittings and hoses were likewise unusable so an entire new system was grafted to the S2 chassis. Starting at the front "T" master was added. New lines and flex hoses to both front and rear hydraulic brakes.

Lotus master cylinder mount to heavy plate inside of "T" section and contains a bell crank attached to the brake pedal. This box had to be rebuilt due to rust.

Also mounted at this time is the steering rack which as is typical was refurbished with cleaning new lubrication and new boot ends. The spline and steering knuckle too were checked before the body is placed. This column traverses the front and rear walls of the "T". Also relocated to the front is the hydraulic switch for the brake lights.

The front suspension was completed by replacing all nuts and bolts, attaching the sway bar and setting a preliminary toe-in adjustment.

Rear brake lines split at "Y" and have flex hose to the swing arm of the rear wheel suspension system.

Further items to be addressed include threading oil pressure gauge line, speedometer cable, emergency brake bell crank and horseshoe balance bracket. Also the chassis blanket sound and heat deadener are draped over the center chassis portion.

Padding installed, bell crank connected (far left) and brake lines at rear to swing arms.

Once assembled many parts need to be displaced for the body to fit such as the master cylinder, e-brake bell crank operator and carburetors to mention a few. Others are parts of cooling system and exhaust muffler. It really would be nice if everything could be left in place in particular the brakes difficult to access.

Another view, must remember to remove all tools and yes those six bolts at the angle brackets, oops forgot before body came. This is now November of 2010 five years into restoration.

Another winter season has come to the Midwest and home of Europa del Sol where for the last several weeks it has resided at the paint shop. Several panic calls from the shop asking for proof of heritage of this strange aberration of a motor-vehicle. All collision shops in Illinois are subject to inspection by the state auditors which requires proof of ownership and/or where cometh all the pieces parts on the premises of said shop for said vehicle. A very technical way of preventing "chop shops" and as such I had to scurry around to locate the title (remember still not titled in Illinois) from Massachusetts and vehicle ID plate somewhere in the various pieces parts yet to be reassembled or even scheduled to be found. Success kept us legal for now, remember the title problems of my Elise, being a salvage title, I now keep any and all scraps of paper pertaining to parentage and purchases involved with many photos of original and current condition for state and insurance purposes.

The "cherry picker" engine hoist was again to place the body and the wheel around cart on the trailer for transport to the paint shop, the contents of the wagon later carried all doors, hoods and miscellaneous parts for their position at the shop. Being partial to orange or other bright colors, as a collision avoidance device, a color of the day was chosen; LO14 or Colorado Orange, an available color in 1970, which may be the only stock item on del Sol.

Being winter it only seemed normal for the scheduled reuniting of the body and chassis take place during a minor snow storm. A reverse procedure was done removing the body from the trailer, trailer backed out and the completed chassis was rolled in under the sky-hooked body. Photos taken look blurry but this is just a deluge of snow christening the whole reunion having been apart for several years. One thing we forgot and created a pain, slightly squished hand were the six center bolts on the chassis these were left in the holes so they would not be lost. Also the locating of the bell crank arm of the hand brake which sets between two layers of glass is a bear to stick into place without a few nicks by the surrounding glass fiber edges.

All of the movement needed to be accomplished in the elements as the low overhead inside the garage does not allow for lift off the trailer and movement once on the cherry picker is to be avoided on an unlevel surface and transition from driveway, to apron into garage level is a three-fer. A most certain formula for disaster; do not drop the freshly painted beauty. It was avoided.

Garage roof and trees loaded with winter white. Top photo shows active members of the transfer crew in hoods, gloves and assorted cold weather gear, lifting del Sol. Center photo chassis under body ready to be lowered. Final photo shows completed reunion back in garage.

"Home at Last, Home at Last, Thank You Lotus Guy, I'm Home at Last".... "Reunited and it feels so good".

Since del Sol has some, (SIC) slight modifications the next step is to adapt the electrical wiring. The original dashboard and controls were trash and a full set of controls, dash and wiring were salvaged from a later model, Europa TC Special. Location of major instruments required a different series of mounting holes at the area below the windshield. The entire wiring loom was modified for these main components. Electric windows were removed, none of the security and seat belt system are used, turn signal and brake lighting altered to remove black box of original Europa loom. A fuse and relay panel for electric fuel pump and radiator cooling fans (2) was added to eliminate the load going thru the ignition switch. See photo below, panel was mounted under dash in passenger side next to wiper motor.

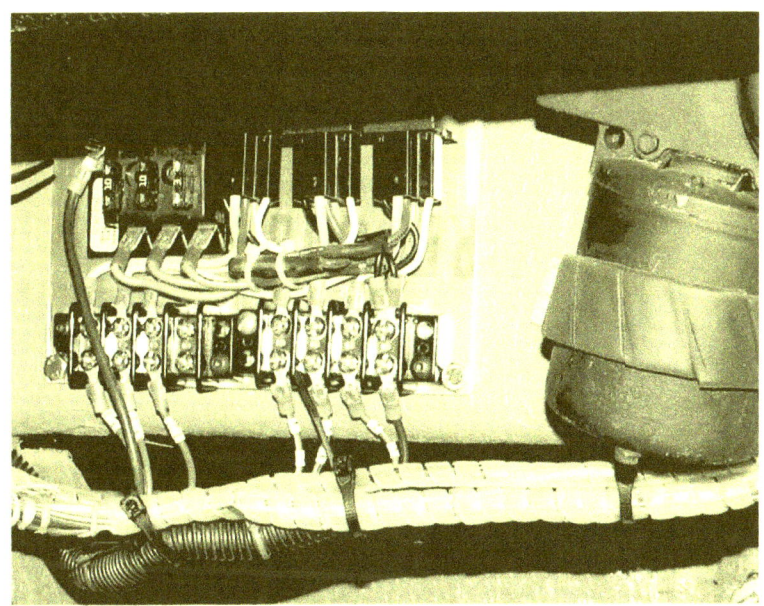

The normal wood dash was modified and covered with carbon fiber, very black and very shinny to complement the door panel inserts also done in carbon fiber.

Once the completed loom is wired and instrumentation installed the dash was affixed to the new, glassed in, mountings. The trick here was to use the two outer holes, unchanged, and install the dash and locate the new aluminum angle brackets.

The photo below shows the new dashboard with switchgear installed for the electric fuel pump, a selection of single or dual cooling fans and manual or thermostat activation of the fans.

Very important to note be sure to complete a diagram of the wiring changes. Because of my electrical background in programming and logic I always lean towards a schematic ladder drawing not the conventional scheme used for automobiles, but since the starting place is the later Europa Federal I started there and as needed drew additional segments, in ladder logic; for the high tech modifications. Also very important to note; any changes where a component is eliminated be sure to fully remove those wires from the loom, leave no hot wires in the loom that do not terminate, floating wires have a way of finding a short to ground, a huge fire hazard potential. Wrapping with electrical tape or (ugh) racers tape is really bad news. Any place you may need to extend a wire; use an insulated crimp connector and shrink tubing to insulate further. Do not just twist the wires together another serious NO-NO!

Final dressing of the wire loom and branches can be fancied up using colorful wrapping tape, made for this application not duct tape, wax cord lacing,(if you know how its done, tricky but I like it, spiral or zip ties come in various sizes and colors. Some use numbers or color coding to distinguish what the wiring is for, i.e. lights, ignition, charging system.

Carbon fiber dash with center switches for fuel and fans. Note the heater controls have been removed as well as the radio, cigarette lighter and window switches, "We don't need those stinking switches," or "Badges" either.

The battery was moved to behind the driver side rear wheel in a reinforced section of the wider rear fender, as the standard location comes too close to the dual Webers.

Wiring and fuel regulator/gauge now traverse this battery evacuated space. And right adjacent to this area is the second 7-1/2 gallon fuel tank. Fuel tanks and batteries should never meet.

Somewhere down the line it will be to my advantage for engine repair or access to fuel tanks, the access via not having a fixed firewall. Remember back at day one it was found the original cardboard and glass wall was trash and totally fell out of the car. The 1" square tube roll cage/stiffener provides a framework with aluminum keepers to hold in an insulating panel in each of the two outside (behind the seats) and the center smaller opening. Following a composite of a fiberglass panel, batting and a black fabric covering to enclose neatly the openings and as these are a very close fit only two grommets along the top edge where the top bracket for the seat belt are affixed thru the steel framework securing the interior panel in place as well.

Insulating fabric was laid over the center tunnel tracing the wiring loom within the padding covered with a second smaller pad and the black felt rug glued in place. Similar rug material was shaped and fitted to the floor and up the door jambs finished off with the standard door seal.

The interior door panels made a trip to Fiberglass Solutions where Paul Q worked his magic in covering the panels in carbon fiber. I finished off all of the panel and sub panels with a black rubber edging and black anodized hardware to screw the panels in place, not using the spring clips normally used. Prior to the panels all door mechanism and latches were redone and checked for fit. No magic here just tons of elbow grease and hours of effort.

The center console was filled in to cover holes associated with the door window switches and ash tray. Graphite paint and still need to replace the gearshift boot.

Before installing the windscreen a fiberglass base was made for the top closure of the dash, padded and layered with the same material on the firewall leatherette. Only drawback was the two defog registers are typically warped and new ones, will, later on be ordered when the final windshield is installed; replacing the delaminating, scratched and fogged existing old glass windscreen.

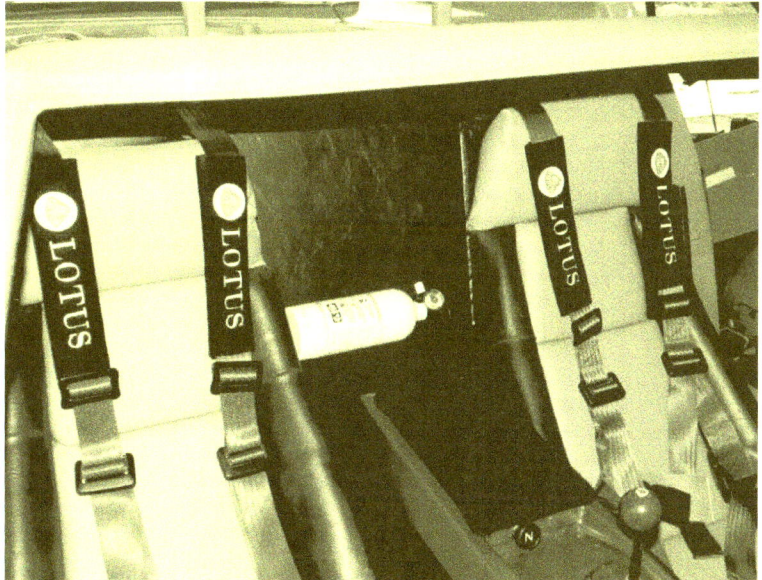

Interiors completed with carbon fiber dash and door inserts, flat dash-board no indents for glove box or gauges. Two tone brown/beige leather seat covering, bright red 4 point seat belts with Lotus announced shoulder pads.

Custom hub for steering wheel to retro fit a 12-1/2"
diameter Formuling wheel. Photo also shows installation of
the clutch and accelerator cables and custom brackets.

Many items of a mechanical nature took place and interspersed with the normal chain of events. One such item was the fuel tanks. The original S2 is a single 7-1/2 gallon tank located on the left side behind the driver's seat/firewall. Later versions Twin Cam/Special had dual tanks; one per side. Both versions inserted the tanks thru an opening in the bottom of the fiberglass body. The existing gas tank was very rusty inside and out no way salvageable including the level gauge sender. Standard tanks are very expensive and custom aluminum or fuel cells likewise. A substitute pair were found on e-bay auction; being auxiliary on large lift trucks (towmotors) . These tanks are steel and heavy gauge, mounting nuts and even had the same capacity of approximately 7-8 gallons, plus had one side angled as the original tanks fitting behind the rakish angle of the seat back and firewall.

Steel brackets supporting the tanks, and bolts on bottom, to be mounted to the reinforced floor of the pocket. Allow the tanks to be installed thru the opening behind the seats. Before installing the tanks were coated inside with an approved liner after applying cleaner and sealer. Exterior surfaces were cleaned primed and painted. The sending units in the tanks were not appropriate having the wrong resistance range. The Lotus units are side mounted and not suitable here so an alternative of a marine unit was applied having both the electrical and a mechanical dial indicator and mounted vertical in the same hole pattern and opening.

A single fuel filler/gas cap was selected with the two tanks connected in a "Tee" with a fuel filter and electric fuel pump both located at the "Y" section of the frame. A fuel pressure gauge and pressure regulator mount to the frame below the dual Webers.

Right side tank in pocket behind passenger seat.

Left side fuel tank, note all steel roll bar at door pocket is encapsulated in fiberglass as it follows door profile.

After paint and unite frame and body features are seen; including fuel tanks, engine (in center opening) and shape of body stiffening roll frame glassed into body panels.

Part of the body strengthening measure several components were added at the bottom of the lower lip of the body. Most noticeable is the stainless steel, polished strip running from front to rear wheel-wells. One portion is the step out of the side air scoop at 1-1/2 inches wide.

Stainless strip, at lower edge of body. Bright polished.

Interior of the rear fender blowout for air scoop it can be shown the addition of aluminum blocks that act as spacers of the lower lip. ¼ inch bolts, stainless steel, of course enter the lower stainless trim strip continue thru the outer shell fiberglass, next the aluminum blocks, then the aluminum plate which goes from wheel well to wheel well bolted and riveted to the inner fiberglass panel.

The photo also shows the replacement lower seat belt bracket forward in the door sill, also attached to the sandwich of stainless strip, fiberglass outer sheet, steel seat belt bracket assembly, aluminum plate, interior fiberglass body shell and finally the lower portion of the steel roll over bracket. Being duplicated on both right and left side and attached to the roll cage a very strong structure is formed.

Interior of air scoop and side panel. A fiberglass sheet is added under the aluminum blocks and at final assembly fiberglass mat and resin were laid over these blocks.

The forward segment of the stainless trim strip is attached with tiny 8-32 screws and nuts thru the strip/fiberglass/aluminum plate and inner fiberglass sandwich. Both front and rear wheel wells are sealed with a closure sheet enclosing all above.

The closure sheet is removable as at time to time it may be necessary to access the door latch striker plate, difficult reach but, doable.

Setting up the suspension and wheel alignment another important handling, plus safety characteristic of the Europa needing to be addressed. One component of the rear wheel alignment controlling the camber is the "lower link" is a non adjustable tube with bushing ends coupling the rear hubs and the bottom center of the transaxle. Two types of adjustable are easy to construct, one a tube with nuts or inserts welded to each end for ½ fine thread or number two my option heavy wall tubing which is directly drilled and tapped one left hand and one right hand to accept rod end to fit the upright and transmission and a locknut on each end to allow adjustment on the car. Photos below show: before and after, then effect on the camber adjustment.

As the adjustment of the lower link was carried out the car was placed on a set of 4 digital scales to as close as possible set a 50/50 front to back and left to right balance of the cars weight. Preliminary set up:

RF----291#                              RR----428#

                    Total Weight –1440#

LF----291#                              LR----430#

Note the lower photo with adjustable link has the tire flat on the ground. Also note the non-stock Spax fully adjustable shock absorbers allow for spring perch adjustment to jack weight and set height of car.

Did you ever notice the one important piece of paper or part needed to continue on with a project is the last place you look or mysteriously fallen into what can only be categorized as a Black Hole. Well one such item is the build sheet of the Gordini R-17 or whatever the 807-13 is. I know that this 1565cc engine is that because of the number plate on the engine. At any rate trying to get a duplicate has proved impossible and going thru each and every book, folder, binder or file loose-leaf or other in the house, garage, upstairs or down the specification sheet is doomed to all eternity Gone!

Why do I care at this late stage, well questions of cam/crankshaft timing, valve overlap and other significant factors effecting performance have reared up. Granted; way back when the engine was torn apart and re-assembled after a good clean up and inspection. Even set up on the completed frame as a test bed ran and deemed to be acceptable to be useable. Now that the car is completed it can be driven and performance is not fully acceptable realizing full well driving a 40 year old resurrected vehicle and comparing to my 05 Elise is sure to show some; let's call them deficiencies. The annual Highland Games which was around the corner, calendar wise, and several miles away distance wise, has been one of my participator car shows and 2011 was to be del Sol's grand unveiling.

Halfway to the show on this warm morning the temperature gauge starting by saying, "Hey look at me, I'm way up here in the stratosphere". So I slowed my super-slab pace to xxxmph and nursed the car, percolating to the show. The question of speed in miles per hour, will be proved later as the gear ratios do not agree with forward momentum by a factor of about 2?

At the show fellow Lotus owners leapt to my rescue with a gallon or more of coolant once the Europa had cooled sufficiently to install the lifeblood fluid. Now is time to test the result and prompted to start the engine to insure a full measure in the radiator system. Click, click, click, no starter action. "What Now del Sol?" Given the best possible attention the starter has given up the ghost.

Where else but the showcase center of the car show to pull out the wrenches and remove the starter after attempts to bring it back to life, it was hopeless as no amount of CPR and TLC would revive the Rhone brick back to life. Out comes a modern day cell phone with internet and WI-FI coupled to a supplier to order a new gear-reduction starter, del Sol was shocked, (not back to life), but to see what advances in communications had transpired since its demise years ago and resurrection today.

With boyish bravado, I got a push start and left in a huff, a big hole where the starter is supposed to reside. I have driven a Europa many more miles than this without a starter from Michigan to Illinois, just don't stop. That of course was the plan until a few miles down the road and copious amounts of smoke erupted from the rear deck lid. I immediately found the right shoulder and killed the ignition, pulled out the extinguisher and opened the deck lid. In all the action, to fix the starter at the show, it appears we, and I do mean me singular not plural, hadn't replaced the radiator cap. All the anti-freeze and water were gone. A quick call back to the show sight brought the Lotus Pit Crew; tow vehicle and trailer to the rescue, whereupon del Sol was carted to a safe storage.

Now back to the reason for build sheet. Once back home and the new starter installed the car was terrible and not running worth a plug nickel. Compression test and leak down showed something was amiss in the power-plant. Too many question would remain un-answered and a full tear down is now required.

Just a note about the starter, a gear reduction unit arrived with a different gear diameter and number of teeth. If installed it would quickly trash the gear and or the ring gear. I quickly checked with the supplier not wanting to make little pieces out of big and was told that they compensate for the size difference, knowing full well the Renault layout, by altering the mounting of the adapter plate. Sure enough once installed no clack or clatter just the sweet purr of the engine spinning upon request, too bad the rest of the engine components would not likewise co=operate.

Twin cooling fans on a VW radiator could not keep del Sol in a normal cool range. Too bad they were fighting Europa outside-in cooling air flow created by negative pressure at deck.

Lotus Corps Pit Crew to save del Sol at Highland Games. Paul's race trailer supplied tools and charger while Fred and Darryl showed knobby knees in kilts.

Europa del Sol's first outing at 2011 Highland Games Car Show was a spectacular disaster. The overheating and starter proved much more needed to be done. It was anticipated that the addition of the side air scoops resulting from the widening of the rear quarter-panels would instill more cooling air into the now rear mounted radiator. Not a significant enough flow to overcome the negative pressure at the rear deck of a Europa. Different means of ducting were tried all failing providing only amounts of aluminum sheet, bent, riveted and otherwise formed to duct air, all ended up in the scrap heap. Original air flow was predicted to enter over the fuel tanks to flow out the four deck lid holes or thru the rear thru the radiator with the assistance of dual fans pulling fresh air out the radiator.

Others have reported in the past this problem so I surrendered and re-located the twin cooling fans to accommodate an air flow in from the rear grill. Also the flow from side air scoops was directed up through the sail area of the fenders over the rear wheel wells and into the rear trunk box. As of the overheating fiasco the engine is not responding properly. Leak-down and compression testing results are poor and timing of cam to crank looks weird. So it is decided to investigate by checking cam timing.

One added problem suspected with the cooling is the lack of an expansion tank on the stock swirl pot. Any over flow out is not replaced automatically causing a continuing fluid loss, more overheating, more fluid displaced more overheating, a vicious circle. A "rat-rod" expansion tank was added, polished stainless steel durable, functional and pretty to boot.

Back to the engine, start tear down, with the removable firewall the exposed front of the engine (timing case) it is not necessary to pull the engine. A suspected valve problem said "pull the head and drop the pan", to remove the cylinders and replace the rings and/or the pistons and liners. After inspection everything was in good shape except the rings. A set were ordered to fit the 77mm hemi set up, also ordered a complete gasket set and new timing chain.

Once back together the valves were reset only I did it wrong in sequence and timing may have been off enough that when the engine was first cranked a clatter unfamiliar came and went very rapidly, It was assumed the distributor was not tied down enough, once re-installed no more noise and of course no operation at all, so after many different trials the valve cover was removed again to double check valve setting and a big OOPS was found. All of the push rods were bent, hitting the pistons with the valves is suspected, Pull the Head again. Yep, all of the valves are no longer straight in particular the Intakes, Bent and forever Useless. Simple order a new set of valves, push rods and rebuild again, Oh yeah, another gaskets set.

Not that simple in addition to a supposed Isky cam of the RG-6A version, (remember I can't find the original build sheet) the valves are not stock or even oversize stock. Instead of a head diameter of 40mm they are 41.5mm or in twin cam terms a Big Valve. These valves are made of the Lotus terminology of "Unobtainium" a very special material. Not even Renault specialist in the area or abroad have seen of late this specification. I did find Datsun valves in head diameter, seat angle and shaft diameter to fit only slightly longer and thought they might work. No can do the shaft length is just too far out of spec. A local machine shop was able to machine the length and relocate the keeper slots to fit and it only cost way, way too much and took three weeks longer than promised, typical.

Since everything was apart it was decided to look into the carburetors. As the car did percolate at high enough and long enough RPM's the possibility of ingesting those coolant fluids into the intake system. Very true as the sickly smell and goo feeling was perceived in the choke area. Further examination of the Weber's found questionable component fitting of different size, mismatched, chokes and later jets in the fuel circuit. Turn the Webers over to Cwik for a complete overhaul.

Trash valves, freshly bent by pistons.

New
valves
installed
in head
ready to
mount
on
bottom
end

Can the experts be wrong; does lightning strike twice in the same spot. After the rebuild of the rebuild; the engine was still balking and timing/ignition/fuel all being in question.

To verify valve/cam/crankshaft timing I installed a makeshift timing wheel on the camshaft pulley (to water pump) and put a gauge on the piston position, intake and exhaust valve of #1 cylinder, the flywheel end of Renault Gordini 1565cc. A Himalayas profile of valves verified close proximity of lift, duration and overlap to the theoretical of the Isky RG-6A profile of: cam lift 0.293, valve lift 0.430 and duration of 258 degrees. No. 1 cylinder piston position at TDC of exhaust/intake stroke confirm exact center overlap of the two valves at TDC is, as that indicated in the Isky Cam Timing missive. All valve lash double checked at .010 and .012 per Renault book specification.

The new Pertronix distributor (replaces the Bosch) was cold set observing a spark plug at 10 degrees BTDC. Plugs and ignition wires were swapped, rotor position checked and a direct lead to the coil bypassing the Tachometer coil and other ignition wiring.

All tests were checked at each step with engine runs, showing no improvement at any of the above tests to produce a smooth idle or reduction of backfires. Even purchased a new Pertronix rated 3.0 Ohm, (chrome plated of course), coil, replacing the Bosch (ugly black).

Early on the compression was checked to verify the new rings had indeed seated in fresh honed cylinder liners, using 3 leg cylinder hone. Each cylinder indicated an excellent reading of 165psi 1-4 which also confirms the valves are likewise working well with good equal seals of the Big Valve, intakes and exhausts.

Surprisingly the engine would run relatively well on choke and at higher revs. As the Webers had just been redone, by an expert, they were deemed okay. Removing a mixture idle screw a gritty-grey substance was noted. Another removal #(?) of the carburetors from the head and dismantling the Webers for another cleaning proved to be The FIX! All passages were flushed with carb cleaner, floats checked and inlet filter screens rinsed though nothing was on the screens. Where the debris came from is undetermined and do not really care. Just certain float chamber and all passages have no present sign of contamination. Some sign of backfire and bucking, minor, were done away with once the idle jets were leaned out one step and the high speed jets were made a single step richer performed by Cwik of course. You see lightning can strike twice in the same spot proving the experts do not always get it right the first time.

Figure below shows the results from the cam test showing in particular the overlap of intake and exhaust valves at TDC of the exhaust stroke. Any deviation either way will have one of the valves contacting a piston at top of the stroke. The Isky manual stresses that overlap gives optimum performance in a very detailed explanation. As the Gordini is only a single cam for both intake and exhaust valves an offset or correction is not possible unless a custom profile cam is configured.

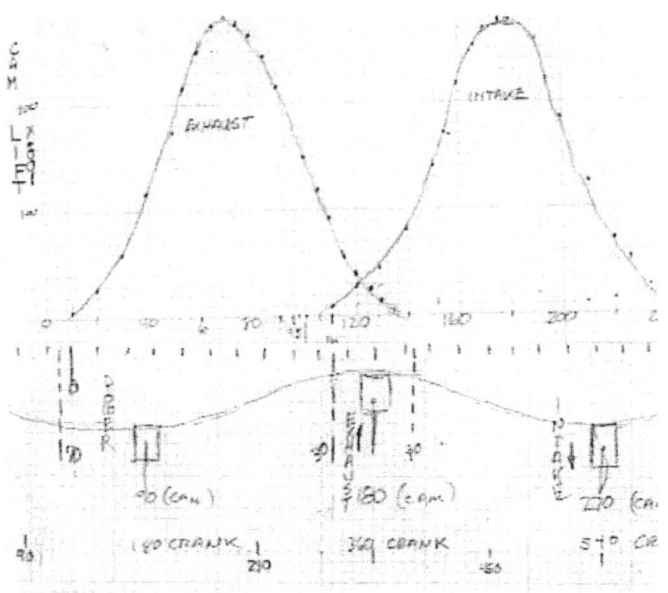

Cam/crank vs. valve position to analyze valve timing.

Cleaned once again: ready to install for the umpteenth time. Weber carburetors DCOE-40

For now and for another day a to-do list of further refinement will wait until a test period of a few hundred miles to insure nothing major needs be addressed.

A final fix? Never.

Good enough for today, we shall see.

*Enjoy the drive*

www.ingramcontent.com/pod-product-compliance
Lightning Source LLC
Chambersburg PA
CBHW051336170526
45166CB00002B/832